THE SUN IS ~~not~~ YELLOW

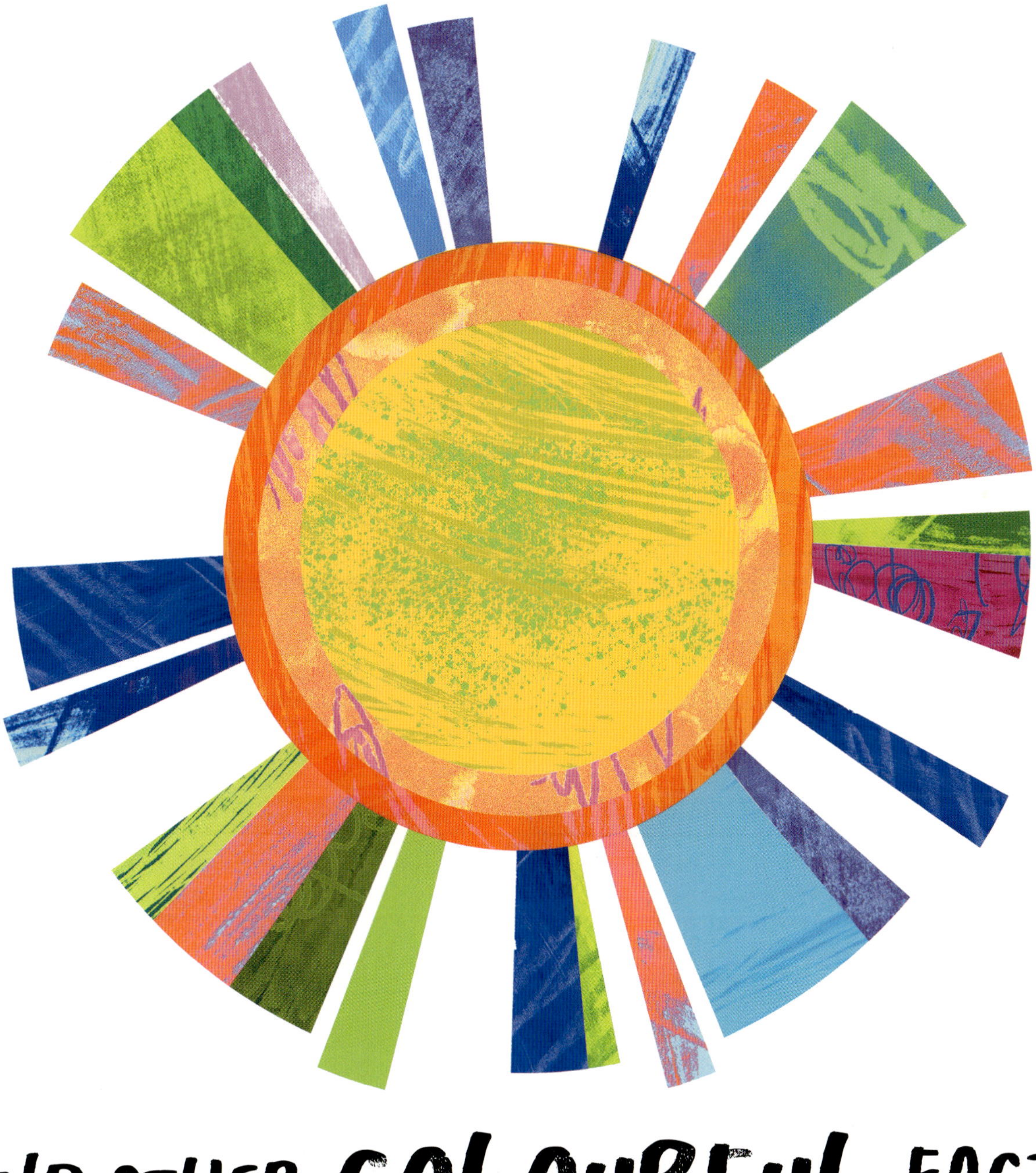

AND OTHER COLOURFUL FACTS

barn owl

Published in 2024
First published in the UK by Barn Owl
An imprint of Igloo Books Ltd
Cottage Farm, NN6 0BJ, UK
Owned by Bonnier Books
Sveavägen 56, Stockholm, Sweden
www.autumnpublishing.co.uk
Copyright © 2024 Barn Owl

All rights reserved. No part of this publication may be
reproduced or transmitted in any form or by any means,
electronic, or mechanical, including photocopying, recording,
or by any information storage and retrieval system,
without permission in writing from the publisher.

1224 001
2 4 6 8 10 9 7 5 3 1
ISBN 978-1-83795-939-6

Written by Cath Ard
Illustrated by Emily Cox and Louis Vettese
Cover Designed by Ashtyn Botterill
Designed by Aimee Swallow
Edited by Katie Taylor

Printed and manufactured in China

At Bonnier Books UK, we are
committed to publishing sustainably.
Find out more here:
bonnierbooks.co.uk/sustainability

THE SUN IS NOT YELLOW

AND OTHER COLOURFUL FACTS

COLOURFUL

CHAPTER 1: Colourful Cosmos 6

Blue Planet? 8

Colourful Cosmos 10

Wonderful Rainbows 12

The Sun Is NOT Yellow 14

Watery Illusions 16

Earth's Most Colourful Places 18

CHAPTER 2: Colourful Nature 20

Look at Me! 22

Stay Away! 24

Colour Transformation 26

Polar Bears Are NOT White 28

Animal Vision 30

Why Is Grass Green? 32

Pretty Amazing 34

CONTENTS

CHAPTER 3: Colourful Human Body 36

Colour Vision 38

Colourful Senses 40

Blue Eyes Are NOT Blue 42

Colours and Feelings 44

Big Questions About the Body 46

CHAPTER 4: Colours and Cultures 48

Eye-Popping Colours 50

Creating Colours 52

Colour Codes 54

Flags of the World 56

Special Colours 58

Colourful Celebrations 60

Colourful Language 62

INDEX 64

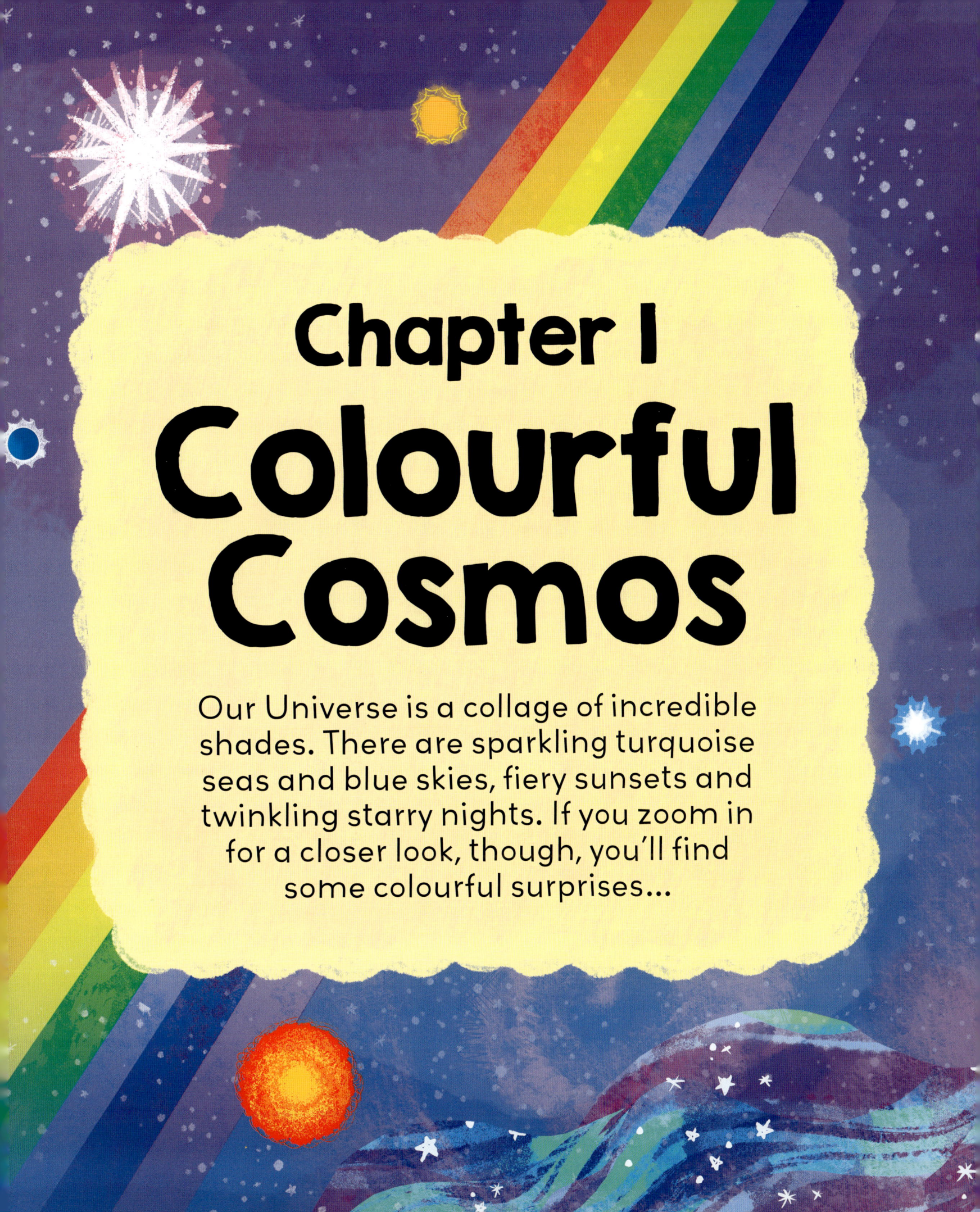

Chapter 1
Colourful Cosmos

Our Universe is a collage of incredible shades. There are sparkling turquoise seas and blue skies, fiery sunsets and twinkling starry nights. If you zoom in for a closer look, though, you'll find some colourful surprises...

BLUE PLANET?

It might be more colourful than you think!

BLUE

Nearly three quarters of the Earth's surface is covered in water, which is how it gets the nickname the "Blue Planet". Almost all of the water is sloshing around in the salty oceans. Just a tiny amount of the water on Earth is fresh water, flowing in streams, rivers and lakes.

GREEN

Just over a quarter of Earth's surface is land. Half of that land is grassland or lush green forest. There are over 3 trillion* trees growing on Earth – that's about 400 trees for you and every single person alive. 12,000 years ago, before people cut down trees to grow crops, there were around 6 trillion trees!

*A trillion is a thousand billion. It is written like this: 1,000,000,000,000.

WHITE

The white caps at the top and bottom of our planet are the frozen polar lands. They shrink in the summer and grow in the winter as the ocean freezes over.

GREY

Bands of wispy grey-white cloud float across Earth. About 2,000 thunderstorms rumble around our planet every minute of the day. The circular clouds are giant, whirling storms called hurricanes. They can produce winds as fast as 306 km/h!

BROWN

Very dry parts of the world look brown from space. Deserts cover a fifth of the land, and can be hot, like the Sahara Desert in Africa, or very cold, like the Gobi Desert in Asia. Only 20 per cent of deserts are covered by sand though; some deserts are rocky and some are even icy!

COLOURFUL COSMOS

When you gaze at a starry night sky, it looks black with specks of white, but if you zoom in, you will discover that space is quite a colourful place!

FAKING IT

Many of the pictures we see of space are taken by powerful space telescopes. They hurtle through space, snapping the Universe and sending back pictures for scientists to study. They show galaxies glowing with incredible colours. However, this isn't exactly what we would see if we explored space...

As well as visible light that our eyes can see, stars give off other invisible rays. The Hubble and James Webb Space Telescopes can spot many of these. Special filters are added to the space images to give the invisible rays a colour that our eyes can see.

This allows scientists to study the images and learn more about space.

MULTICOLOURED PLANETS

- There are seven other planets that travel around the Sun with Earth, and they all look very different!
- Mercury is solid, dark grey rock.
- Venus has swirling clouds of yellow and white gas.
- Mars is covered in rusty red rocks.
- Jupiter has swirling clouds of yellow, orange, brown and white gas.
- Saturn is surrounded by rings of brown and yellow rock and dust.
- Uranus gets its milky blue hue from white clouds and methane gas.
- Neptune's deep blue colour comes from methane gas. (Unlike Uranus, it doesn't have many clouds to make it look lighter.)

YOUNG AND OLD

Stars live for billions of years. As they go through their long life, they change colour, size and temperature.

A yellow dwarf star is an average star, about the same size as our Sun. It swells and cools as it gets older and becomes a red giant. Eventually, it dies and becomes a tiny white dwarf.

Supergiant stars are far bigger than our Sun. As they get older, they swell and cool and become red supergiants. Eventually, they die in a spectacular supernova explosion. All that remains is a tiny, spinning neutron star. When an extra-massive star dies, it forms a black hole.

STARS ARE NOT WHITE

Stars are giant balls of exploding gas that shine with a range of colours. The Sun is our nearest star, but most stars are billions or trillions of miles away in space. The light from a distant star is so faint that our eyes cannot detect any colour – it just looks white.

Blue or white stars are massive, hot and bright.

Yellow stars are average-sized and medium hot.

Orange and red stars are generally cooler and dimmer.

RED IS NOT (ALWAYS) HOT

Red usually means hot, and blue cold, but with stars it's the opposite! There is no such thing as a cold star - all stars are unbelievably hot - but some are much hotter than others. Hotter stars are brighter while cooler stars are dimmer.

WONDERFUL RAINBOWS

On a rain-and-shine day, a rainbow can magically appear in the sky. But how do raindrops and light create a band of beautiful colours?

This bit is the prism!

SOLVING THE MYSTERY

For thousands of years, people tried to figure out what makes a rainbow. In 1672, English scientist Sir Isaac Newton set out to prove that the colours come from white light. He took a triangular piece of glass called a prism and shone a beam of light through it. The light bounced off the inside of the prism and spread out into seven colours: red, orange, yellow, green, blue, indigo and violet.

Next, Newton directed the coloured light back through another prism to turn the rainbow back into a beam of white light again. This proved that the colours come from the light, not the prism!

BENDY, BOUNCY LIGHT

Every raindrop in the sky acts like a tiny prism. When it travels through water, light bends, then bounces back off the inside of the raindrop and scatters into its different colours.

The order of the colours never changes. Red is always at the top and violet is always at the bottom. This is because the colours in light bend by different amounts and bounce off the raindrop at different angles.

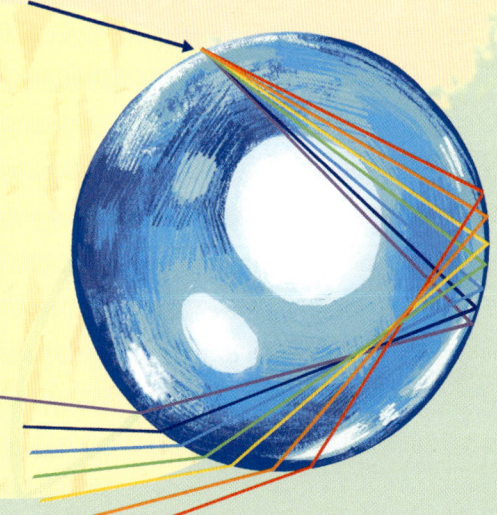

Bouncing beams

YOUR OWN RAINBOW

You have to be in just the right place to see a rainbow, with the sun behind you and rain falling in front of you. Someone in a different place will see a different rainbow, or they might not see one at all!

REMEMBER THE RAINBOW

Some places have special ways to remember the seven colours of the rainbow. For example, in the UK, they say "Richard Of York Gave Battle In Vain", where each word starts with the same letter as the colours in the rainbow. (This is called a mnemonic device.) In the USA, they say "ROY G. BIV", and Russian-speakers say "Every Hunter Wants To Know Where The Pheasant Sits".

RAINBOW-TASTIC!

Rainbows can form wherever there are water droplets in the air!

- The sun shining through spray from a waterfall or fountain can make a rainbow.
- Pale moonbows appear when moonlight travels through rain or mist.
- When a whale blasts misty air out of its blowhole, it can make a blowbow!

GUESS WHAT?

Raindrops don't look like teardrops. Small ones are ball-shaped whereas bigger ones are like hamburger buns. Really big raindrops are like jellybeans.

TRY IT!

You will need:
- Sunshine (it's best when the sun is low in the sky)
- A garden hose or a plant sprayer bottle

RECIPE FOR A RAINBOW

1. Stand outside with the sun behind you.
2. Spray a mist of water in front of you.
3. Sunlight bounces off the inside of the water droplets and you see a rainbow!

THE SUN IS NOT YELLOW

We see brilliant blue sky and golden rays of sunshine, but it's all a trick of the light. Not only is the Sun NOT yellow, but the sky is NOT really blue!

To see the Sun's true colour, just look at the Moon. We see it as white in the night sky, but it actually only looks that way because it is reflecting the Sun's white light.

SCATTERING LIGHT

A layer of gases called the atmosphere surrounds Earth. The gas molecules scatter the sunlight as it travels through. Blue and violet are easily scattered because they travel in shorter, smaller waves. This is especially true for ultraviolet light, the purpliest of the purple, which humans can't even see!

Whichever way you look, you see the scattered blue light, which is what makes the sky appear blue. With fewer blue wavelengths reaching our eyes, the Sun appears yellow.

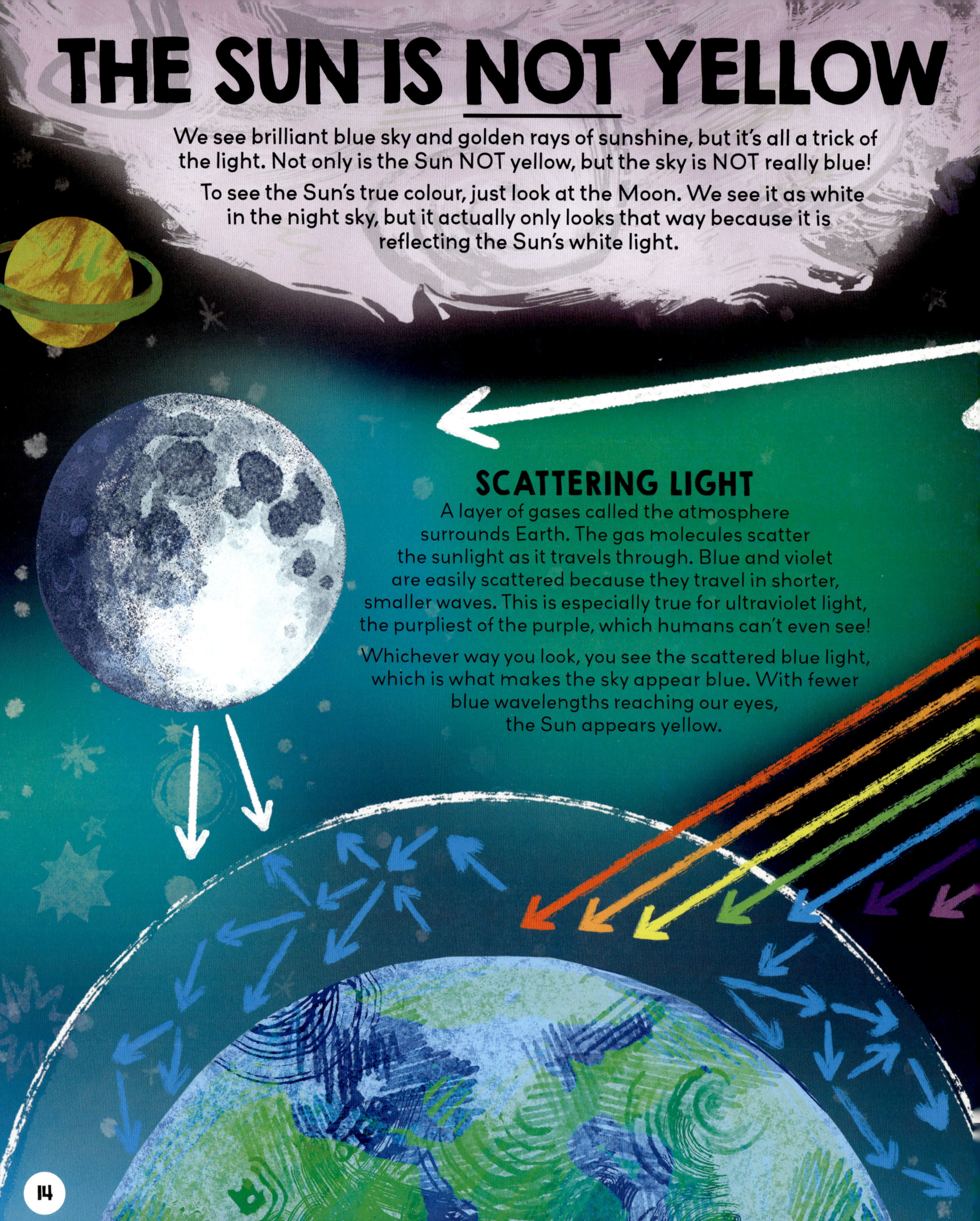

TRAVELLING LIGHT

The Sun is white, and white light is made up of all the colours of the rainbow.

Light travels in wiggly waves. These come in various wavelengths, with each one representing a different colour.

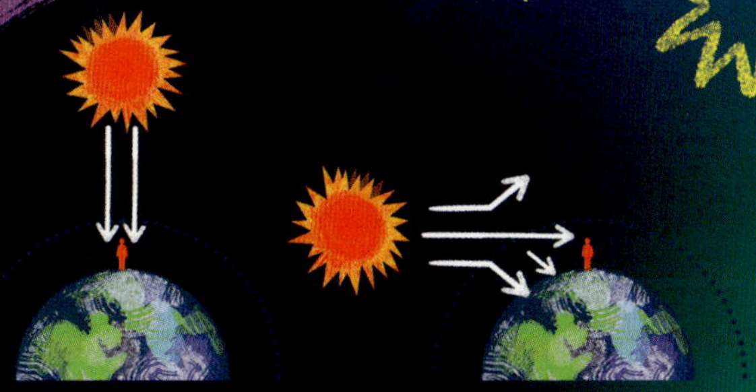

WHY ARE SUNSETS RED?

At sunrise and sunset, when the Sun is low on the horizon, its light has to travel further through Earth's atmosphere. As well as gases, the dust, pollen and water droplets lower in the atmosphere scatter the blue and violet light, leaving red and orange rays to reach our eyes.

TRY IT!

You will need:
- A straight-sided glass
- Water
- Milk
- A teaspoon
- A torch

SUNSET IN A GLASS

1. Fill the glass two-thirds full of water and add a teaspoon of milk. Give it a stir.
2. Make the room as dark as possible.
3. Shine the torch onto the side of the glass and look from the opposite side. The milk has a red tint!

Torchlight is white, like sunlight. The milk particles in the water act like the particles in the atmosphere, blocking the blue and green waves of light so that we see only the orange and red.

WATERY ILLUSIONS

Have you ever wondered why water can appear white or grey in the sky, and blue in the ocean, when it's actually clear? It's all down to the fact that many of the colours we see in the natural world are actually light-scattering illusions!

WHY ARE CLOUDS WHITE?

Clouds are made up of transparent water droplets. When sunlight travels through these droplets, its colours are all scattered equally. This means the white sunlight remains white and the clouds appear white too.

WHY ARE RAINCLOUDS GREY?

Rainclouds are densely packed with water droplets. They are also thicker, heavier and taller than other clouds. The large droplets scatter more light, but the dense cloud means that less light gets through to the base of the cloud, so it looks dark.

DID YOU KNOW?

The tops of clouds are always white because they are bathed in sunlight. Look out of an aeroplane window when you're above the clouds and you'll see brilliant white, even if it's raining down below.

WHY IS THE OCEAN BLUE?

It's not due to it reflecting the blue sky! The ocean looks blue because water absorbs red, yellow, and green light while scattering blue.

The more water molecules there are, the more light particles will collide with them and be scattered, making deeper water appear a deeper shade of blue.

EARTH'S MOST COLOURFUL PLACES

Put on your shades and take a colourful tour of some of the most dazzling places on Earth.

BURNING BLUE
Take a night hike to the Kawah Ijen volcano in Java, Indonesia. It looks like electric blue fire is flowing from the crater, but the eerie glow is actually caused by burning sulphuric gases.

SURPRISING SPRING
Gaze at the Grand Prismatic Spring in the USA's Yellowstone National Park. Its bands of orange, yellow and green are created by trillions of microscopic bugs living in the water!

RAINBOW RIVER
In the autumn, Colombia gains a liquid rainbow in the form of the Caño Cristales River, when its green moss clashes with the crimson plants that cover the rocky riverbed.

DANCING LIGHTS

Head to the icy Arctic to see wavy green lights dancing in the sky. The Northern Lights are created by storms on the surface of the Sun. These send out clouds of electrical particles that collide with Earth's atmosphere. You'll find the Southern Lights over the Southern Ocean!

MULTICOLOURED MOUNTAINS

The colourful rock formations in China's Zhangye National Geopark look like they have been painted. They were formed by layers of sandstone and minerals being laid one on top of the other over millions of years.

ROSY LAKE

Take a dip in Senegal's super-salty Lake Retba! Microscopic, salt-loving algae that live in the water produce the salmon-pink colour.

BLUSHING BEACH

Sink your toes into the powdery pink sand at Pink Sand Beach in the Bahamas. The rosy tint is created by ground-down coral and the red shells of tiny sea creatures.

Chapter 2
Colourful Nature

The animal kingdom is a rainbow of vibrant feathers and fur, scales and skin. There is a reason for all of the incredible colours that we see in the natural world. Finding a mate, finding food… it can even be life or death!

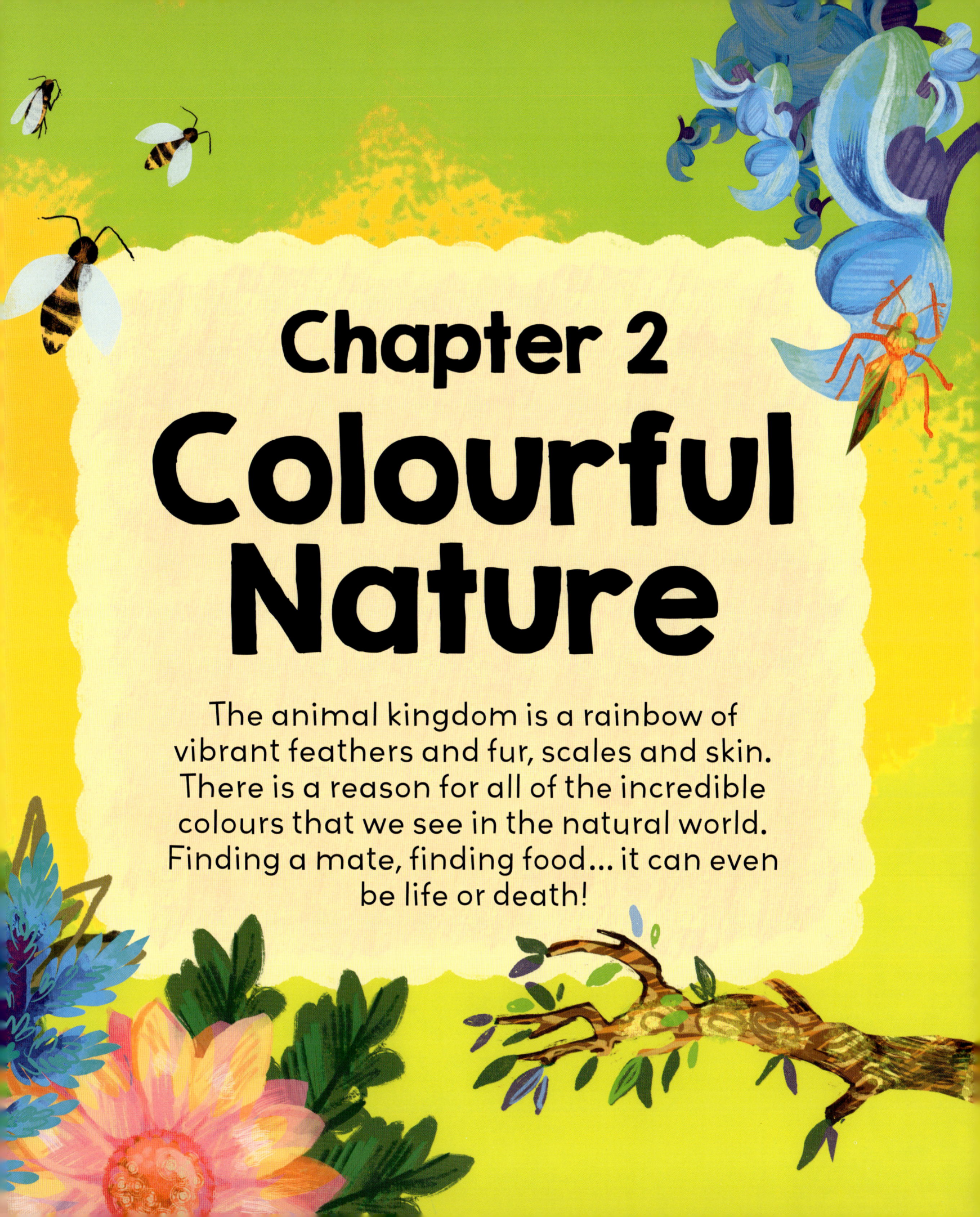

LOOK AT ME!

Attention-grabbing colours can be a way to charm a mate or tempt prey. If colour is nature's way of communicating, what are these animals saying?

I AM THE GREATEST

The mandrill is the world's most colourful monkey, and males become more vibrant as they grow older. The most successful males are the biggest and brightest so other mandrills can easily tell who's the leader.

I'M GOOD, THANKS!

This small, red-faced monkey from the Amazon rainforest isn't blushing or angry. Instead, scientists think the bald uakari has scarlet skin to show other uakaris that it is strong and healthy. A pale face is a sign that a monkey is sick, so the redder the better!

NOW YOU SEE ME...

To attract a mate, the blue morpho butterfly spreads its shimmering blue wings. To hide from danger, it closes its wings and the brown underside blends in with the branches.

SURPRISE!

The male peacock spider might be tiny, but it puts on a big display. When it spots a female, it unfolds a special flap to reveal its multicoloured body. It struts back and forth, flashing its brilliant behind and waving its legs.

FEED ME!

Baby birds beg their parents for food by chirping and opening wide. Their gaping red or orange mouths outlined in yellow or white make an unmissable target for food to be dropped in.

Gouldian finch chicks have luminous spots at the sides of their open mouths that act like a bright beacon to guide parents to them in their dark nests.

FLAPPING FANTASTIC!

The male fan-throated lizard unfurls a flap of skin on its throat, called a dewlap, to reveal dazzling scales. Displaying its jazzy dewlap impresses nearby lady lizards and tells rival males to keep away.

TAKE A LOOK...

The orb-weaving spiny spider is pretty deadly. Its brightly coloured back lures unsuspecting insects. Studies show that the most colourful spiders catch the most prey!

... IF YOU DARE!

The five-lined skink is an escape artist. If a snake or bird bites its colourful tail, it breaks off and keeps wiggling while the skink escapes to grow a new one!

MALE VS. FEMALE

Male birds are often much more colourful than females. This is because it is the male's job to see off rivals and attract a mate. The female's dull colours keep her hidden when she is sitting on a nest of eggs.

STAY AWAY!

Creatures that don't have shells, claws or sharp teeth have to find other ways to defend themselves. Colourful creatures look attractive to us, but to other animals they send a serious message.

Some animals wear their warning colours all the time, while others only show them off when they are alarmed.

Have you realised how many warning signs in our world are red (or yellow) and black? Humans have copied nature, where combinations of black, white, red, blue and yellow are a colour code that spells danger. When we see them, we take notice!

When a skunk raises its tail and shows its black and white behind, it's time to run! This stripy message means it's about to squirt a stinky liquid.

The oriental fire-bellied toad blends into its mossy home – until it feels threatened. Then it arches its back or flips over to reveal its red and black belly – a sign that it is packed with poison.

This fuzzy insect is actually a wasp. It looks cuddly, but the female red velvet ant gives a nasty bite.

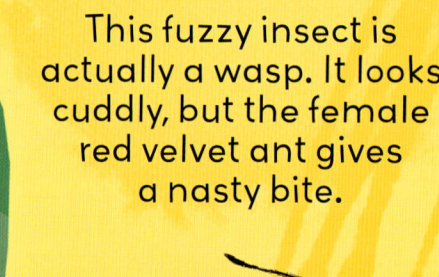

The highly poisonous blue-ringed octopus can turn on its own warning sign if it senses danger by making its vibrant rings glow against the yellow and black of its body.

Deadly Coral Snake

Some harmless animals copy the warning colours and patterns of their dangerous doubles to stay safe. Can you spot the difference?

Poison takes energy to make, so some animals choose the easy route and steal it from their food!

The day-glow shades of many sea slugs basically scream, "Do not eat"! The slugs hunt stinging jellyfish and sea anemones, then use the toxins to make their own weapon: poisonous slime.

The monarch caterpillar's colourful body is a clue that it is not a tasty snack. The milkweed plant that the caterpillar feeds on gives it a disgusting flavour.

Poison dart frogs come in a range of eye-popping colours, which tell predators they are deadly. They eat poisonous beetles, then store the toxins in their own colourful skin.

A flash of the 88 butterfly's eye-boggling wings are enough to dazzle and confuse birds.

Sometimes a flash of colour is all it takes to make a predator think twice.

A frilled lizard can make itself seem big and scary by flaring the orange ring of skin around its neck.

The blue-tongued skink is shockingly rude. It sticks out its bright blue tongue to give attackers a fright.

← Harmless King Snake

COLOUR TRANSFORMATION

Some amazing animals can change their colour to match the season, the occasion or even their mood!

COLOURFUL LANGUAGE

Some animals show their emotions through their skin.

- The panther chameleon turns from a calm green to a hot red when it's fired up and ready to face a rival.

- An angry octopus will make itself darker to appear threatening, while a scared one turns paler.

- The dwarf cuttlefish can change its colour and pattern in a flash. It uses colour to communicate with other cuttlefish, as well as to sneak up on prey.

DRESS TO IMPRESS

It's not just humans who dress up to go on a date!

- A golden tortoise beetle can quickly change its wing casings from gold to red, which is useful for seeing off attackers and impressing females.

- Male puffins are famous for their colourful bills and bright feet, but this look is just for summer breeding. In winter they fade to a much duller colour.

- Scarlet tanagers are only scarlet during the summer when they're looking for a mate. In winter, they fly south to the rainforest, where they change to a dull green to blend in.

DID YOU KNOW?

Carotenoids are pigments (colours) found in tomatoes, watermelon, carrots and lots of other foods. They're also found in the shrimp eaten by flamingos, which is what turns their feathers pink.

HIDE AND SEEK

Changing colour to match the surroundings makes it easier to hide or to hunt.

- Wherever a peacock flounder settles, it can change its skin to match the seabed.
- An Arctic fox changes its fur with the seasons. The fox is an earthy brown in summer, but by winter, it's as white as the snow-covered ground.
- Female goldenrod crab spiders ambush the insects that come to feed on flowers. The tiny spiders can slowly change colour to match the petals of their lair.

BABY COLOURS

Some baby animals start out a very different colour to their parents!

Baby silvered leaf monkeys have bright orange fur. This helps the grown-ups in the group keep track of little ones in the dense jungle.

Flamingo chicks start life with white or grey feathers. Over the years, they turn rosy pink from all the carotenoids in their food.

A zebra shark looks more like a leopard! It gets its name because its young are born with stripes, which slowly change to spots as they grow.

POLAR BEARS ARE NOT WHITE

Polar bears definitely look white, but there's much more to their snowy fur than meets the eye!

Small tail

The polar bear's skin is black, as having dark skin keeps the bear warmer.

A polar bear's coat keeps it camouflaged (and warm!) in its icy Arctic home. It sounds straightforward, but look closer and you'll see that the fur is *not* white – it's all a trick of the light.

Polar bears have a double layer of fur. The long guard hairs are clear, hollow tubes that reflect and scatter light. Because sunlight is white, the fur appears white.

- Guard hair
- Dense undercoat
- Dark, thick skin
- Thick fat layer

Small ears

The hollow guard hairs fill with water when the bear is swimming. The bear's body heats the water to keep itself warm, a bit like a wetsuit. On land, the hollow hairs transfer the Sun's energy down to the polar bear's skin.

TRY IT!

You will need:
- A piece of white paper
- A piece of black paper
- 2 ice cubes
- Sunshine

BLACK AND WHITE

1. Place an ice cube on each piece of paper and place them in a sunny spot.
2. See how long they take to melt.

WHAT'S HAPPENING?

Black absorbs more sunlight, which turns into heat energy, making the paper warm up and the ice melt faster. White reflects the sunlight, so the ice melts more slowly. If you want to stay cool on a hot day, wear white. If you want to stay warm on a cold, sunny day, wear black. The polar bear's clever coat manages to keep the bear both camouflaged and warm!

ANIMAL VISION

Nature is a technicolour feast for the eyes, bursting with brilliant colours and patterns, but very few animals see the world like we do!

WHY ARE TIGERS SO BRIGHT?

The zany stripes of a tiger look like a glowing, hi-vis jacket. However, many animals can't tell the difference between oranges, yellows and greens. To deer and other tiger prey, the cat's orange coat blends in with the browns and greens of the jungle.

 v

WHY DO BULLS HATE RED?

We all know that matadors wave a red cape and the bull charges at it, but did you know that the bulls are not actually enraged by red? In fact, they can't even see it! Like all cattle, they are colour blind – it's just the movement of the cape that makes them mad.

 v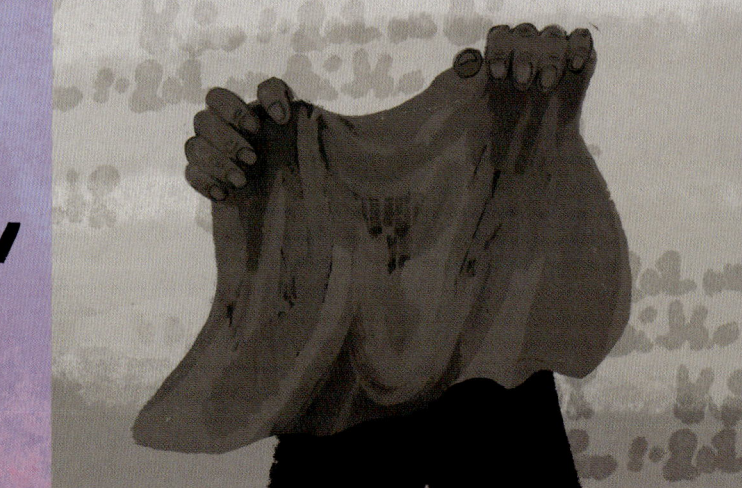

ULTRA EYESIGHT

Bees and butterflies can see ultraviolet light: a kind of purple light that is invisible to us. When they look at a flower, they see secret markings that act like a target, helping them make a beeline for the sweet nectar in the centre.

WHY DO BEEKEEPERS WEAR WHITE?

To avoid being stung! Many animal honey-thieves, like bears and skunks, have dark fur. When bees spot a big, dark shape, they sense danger and go on the attack. However, if you wear light, bright colours, bees and insects are more likely to simply land on you. Don't panic, though; they've probably just mistaken you for a big flower.

NIGHT SIGHT

Having excellent colour vision in low light is important for animals that hunt at night. This is why geckos don't have any trouble tracking down bugs for dinner after dark!

WHY IS GRASS GREEN?

Even in a city, you can probably find some green growing somewhere – a patch of grass, weeds pushing up through the pavement, leafy trees... Green is the undisputed colour of nature, but why?

Living things are made up of incredibly tiny units called "cells". Each grass cell is filled with small parts called chloroplasts, which contain chlorophyll.

Chlorophyll is activated by absorbing sunlight. It changes water from the soil and carbon dioxide (a gas in the air) into sugars, which feed the plant.

INSIDE THE FOOD FACTORY

If you look very closely at a blade of grass, you'll see millions of microscopic cells that act like food factories for the plant. These contain green chlorophyll, which is where the food-manufacturing takes place.

WHY DO LEAVES CHANGE COLOUR?

In autumn, the days get shorter and there is less sunlight. Less sunlight makes leaves stop making chlorophyll. Without the green pigment, yellow and orange pigments in the leaves are revealed. Veins connecting leaves to trees start to close, trapping sugars, which make some leaves, like maple, turn bright red.

GUESS WHAT?

Scientists have shown that being in green spaces is good for our health. It makes us less stressed and gives our bodies a boost to fight bugs.

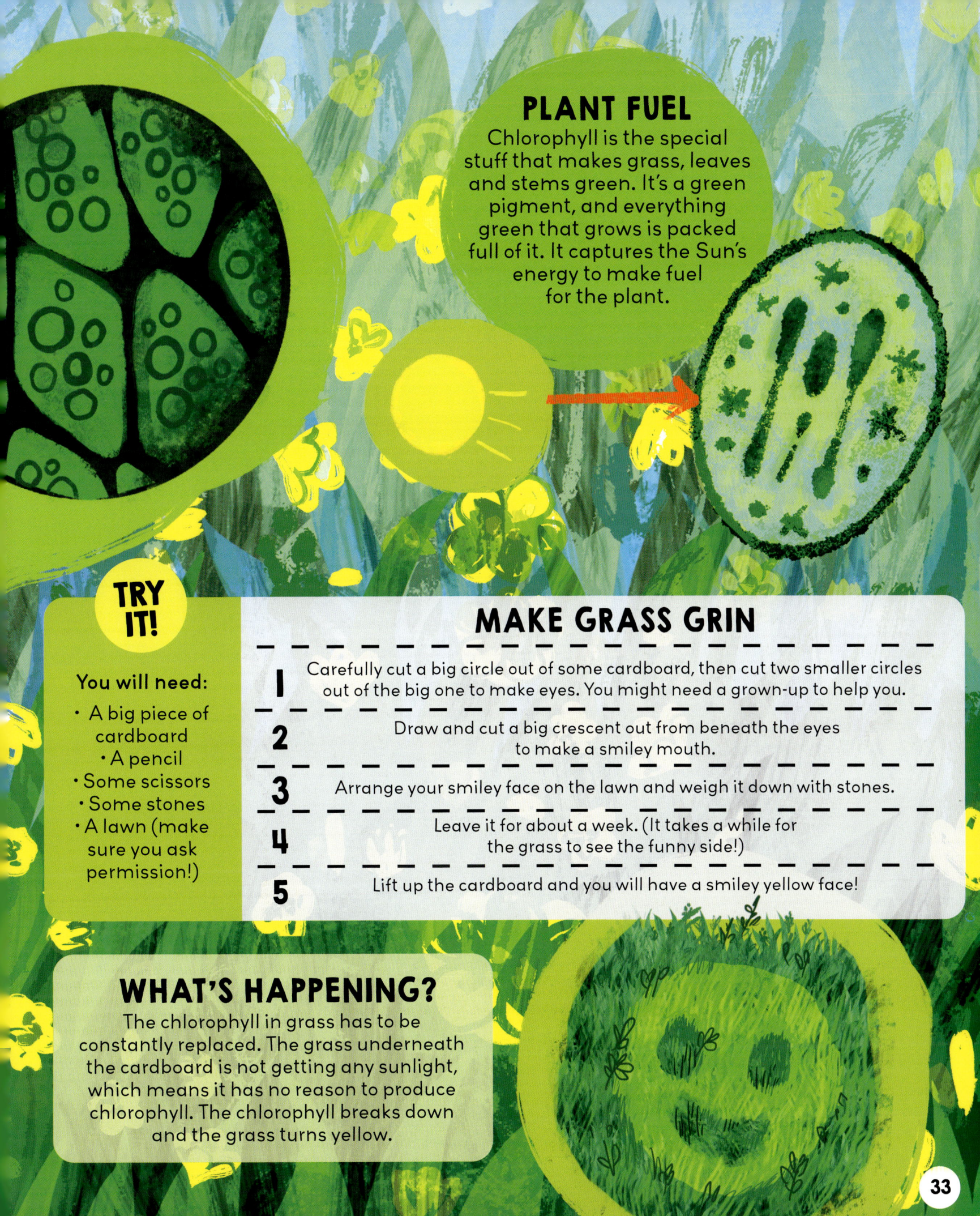

PLANT FUEL

Chlorophyll is the special stuff that makes grass, leaves and stems green. It's a green pigment, and everything green that grows is packed full of it. It captures the Sun's energy to make fuel for the plant.

TRY IT!

You will need:
- A big piece of cardboard
- A pencil
- Some scissors
- Some stones
- A lawn (make sure you ask permission!)

MAKE GRASS GRIN

1. Carefully cut a big circle out of some cardboard, then cut two smaller circles out of the big one to make eyes. You might need a grown-up to help you.
2. Draw and cut a big crescent out from beneath the eyes to make a smiley mouth.
3. Arrange your smiley face on the lawn and weigh it down with stones.
4. Leave it for about a week. (It takes a while for the grass to see the funny side!)
5. Lift up the cardboard and you will have a smiley yellow face!

WHAT'S HAPPENING?

The chlorophyll in grass has to be constantly replaced. The grass underneath the cardboard is not getting any sunlight, which means it has no reason to produce chlorophyll. The chlorophyll breaks down and the grass turns yellow.

PRETTY AMAZING

Plants can't move or make a noise to attract attention, but they've found their own clever ways to get noticed.

BERRY DELICIOUS
Some plants produce berries full of seeds. When they are ripe, they turn a bright colour so that passing birds spot them. After they've filled up, the birds fly off to a new location, where the seeds pass out in their poo and new plants can grow.

GUESS WHAT?
There are 280,000 known kinds of flowering plants on Earth, but only 10 per cent of them are blue!

WATCH OUT!
Wild berries are good for birds to eat, but many are poisonous to people. You should only eat them if you know for sure what they are.

CREATING A BUZZ
Plants need flying insects like bees, butterflies and moths to carry their pollen from flower to flower to make seeds. This is called pollination and is how plants grow new plants. Flowers need to stand out against their green leaves to get noticed – the brighter the better when it comes to pulling in passing insects!

BED AND BREAKFAST

Can you see how the incredible lobster claw plant got its name? It grows in the rainforests of South America where it attracts hungry hummingbirds that drink its sweet nectar. These tiny birds also make their nests in the plant!

RARE BEAUTY

The rare Jade vine is found in the rainforests of the Philippines. Its turquoise flowers droop down to be pollinated by bats that hang from the trees. At night, the flowers have a luminous glow that attracts the night feeders in the dark forest.

FRESH FLOWERS

Some flowers change colour to show insects which flowers are old, and which ones are new and full of sweet nectar. The cotton rose, for example, starts out white and turns pink throughout its life.

TURN TO STONE

Living stones are hard to spot amongst the pebbly ground where they grow in South Africa. Their clever camouflage stops thirsty animals from munching the juicy plant before it has time to flower.

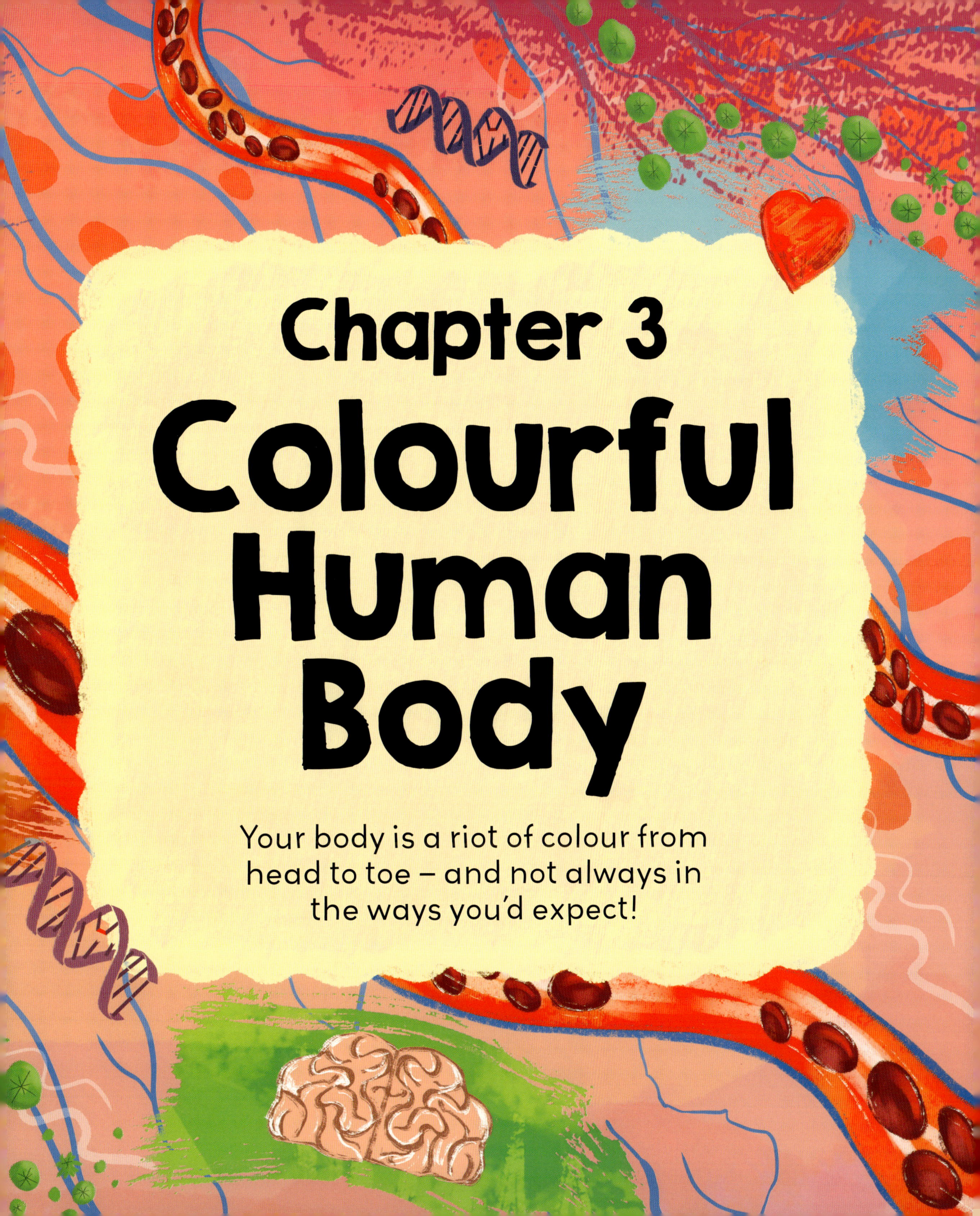

Chapter 3
Colourful Human Body

Your body is a riot of colour from head to toe – and not always in the ways you'd expect!

COLOUR VISION

From the moment we wake up in the morning, our eyes and brain are busy detecting the colours all around us.

SEEING COLOURS

White light is made up of a rainbow, or spectrum, of colours. Surfaces reflect and absorb light differently, which the human eye and brain convert into colour.

When white light strikes a white object, it appears white to us because it absorbs no colour and reflects all colour equally.

A black object absorbs all colours equally and reflects none, so it looks black to us.

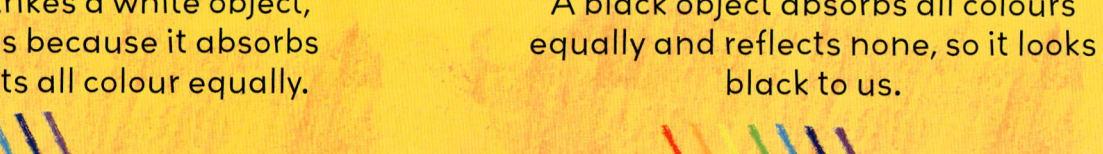

Whichever colour(s) aren't absorbed by an object are reflected back to our eyes.

DID YOU KNOW?

Most people can see more than one million different colours.

COLOURFUL CONES

We have light-sensitive cone cells and rod cells at the back of our eyes, which send messages to the brain. Cone cells detect colour. Some detect red light, some detect blue light and some detect green light.

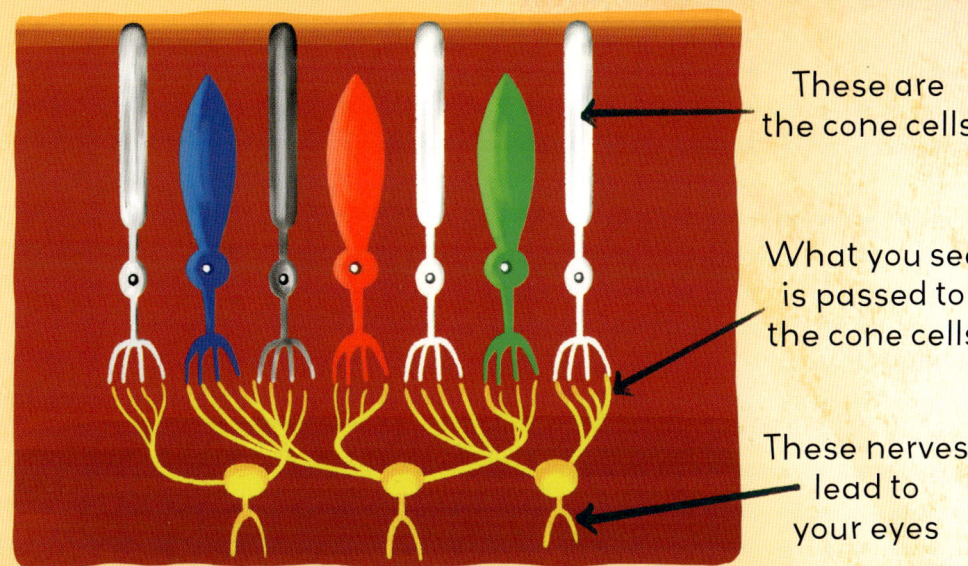

These are the cone cells

What you see is passed to the cone cells

These nerves lead to your eyes

The brain then converts these signals into red, blue and green – and all the many colours in between.

NIGHT VISION

Cone cells only work in bright light. To see in the dark, our rod cells take over. They can see the shape of things, but they don't detect colour. That's why we see everything in shades of grey at night.

TRY IT!

MAGIC BIRD

Stare at the yellow canary for 15–30 seconds, then move your gaze to the white square. Does a blue bird appear?

WHAT'S HAPPENING?

Red and green light mixed together make yellow. The red and green cone cells have had to work hard sending signals about the yellow bird to your brain. They take a break while the blue cells take over for a few seconds. This makes you see an "afterimage" of the bird in blue.

COLOUR BLINDNESS

Some people have two types of cone cells in their eyes instead of three, so they see the world differently.

Normal vision (Three cones)   **Deuteranomaly** (No green cone cells)   **Protanomaly** (No red cone cells)   **Tritanomaly** (No blue cone cells)

 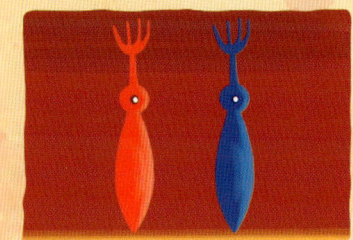 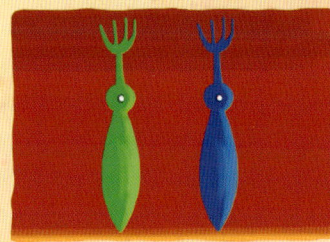 

39

COLOURFUL SENSES

How food looks plays a big part in its flavour. We don't just taste it with our tongue; our eyes and brain are feeding us information before we even take a mouthful...

TASTY COLOURS

Our senses work as a team when we eat. We see and smell food, then feel its texture in our mouth. Often, the brain is telling us what flavour to expect because of past experience: we know what orange juice looks and tastes like, so we prepare for an orangey flavour.

A food's colour might mean we detect flavours that aren't actually there. Most people associate these colours with these flavours...

Red and pink = sweet or spicy

Yellow and green = sour

White = salty

Brown and black = bitter

MMM! SMELLS GREEN!

Some people have a super-sense called synaesthesia, where two or more senses link up.

Colours might have a taste, or different words might have their own colour!

GROSS GRUB

Food can taste weird or nasty if it is a different colour from what we expect, especially if it is green or blue (the colours of mould). It's our body's way of protecting us from eating gone-off grub.

FROM YUM TO YUCK!

In one scientific study, volunteers were given a dinner under lighting that hid the food's true colours. The sneaky lights were switched off and a blue steak and green chips were suddenly revealed. Some people were actually sick!

TRY IT!

You will need:
- 3 fruit juice (or squash) flavours
- Cups
- Red, blue and yellow food colouring
- Some volunteers

TASTE TRICKS

1. Put one flavour of juice in three different cups and colour them all differently with a few drops of food colouring.
2. Ask a volunteer to sip each one and describe what they taste.
3. Try it again, putting a different flavour juice in each cup, and dye them all the same colour.

WHAT'S HAPPENING?

In the first experiment, the drinks taste exactly the same, but people might detect different flavours because of the colour. In the second experiment, the drinks taste different, but people may think they taste the same because they *look* the same.

BLUE EYES ARE NOT BLUE!

Human eyes come in many colours, but if you are reading this with blue eyes then prepare to be surprised: they aren't really blue, and you're *sort of* a mutant…

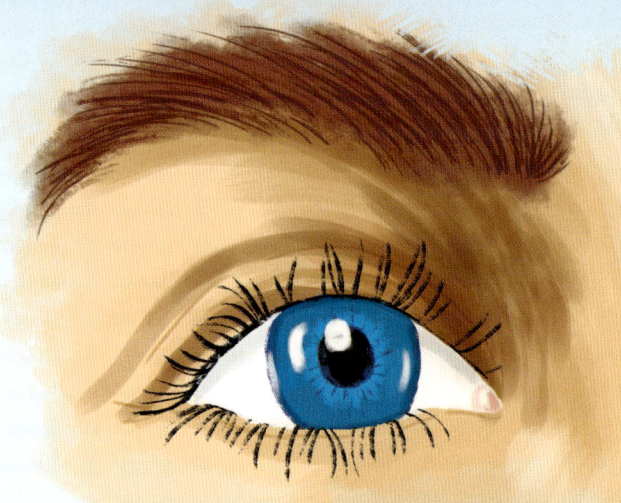

TRUE COLOURS

A pigment called melanin is the stuff that gives our hair, eyes and skin its colour. Melanin is brown. The more melanin you have, the darker and browner your eyes, skin or hair will be.

BLUE-EYED MUTANTS

In the beginning, all humans had brown eyes, then one person changed the world with their mutant genes!

Eye colour is determined by the amount of melanin in the front layer of the iris. Melanocytes are the cells that make melanin pigment.

If there's a lot of melanin, you have brown eyes.

If there's less melanin, you have green or hazel eyes.

If there's very little melanin, you have blue eyes.

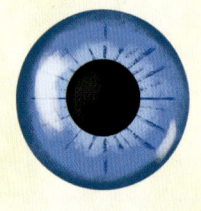
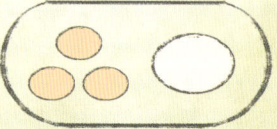
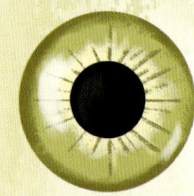
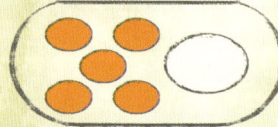
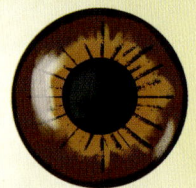
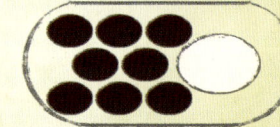

People with blue eyes don't have much pigment in their eyes. The front of the iris is colourless, but it scatters light and reflects blue light back, making the iris appear blue. It's the same reason the sky and water look blue. (Find out about that on page 16!)

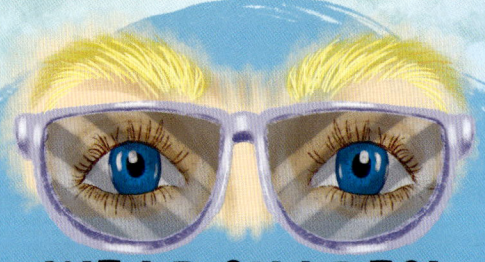

WEAR SHADES!

Melanin protects us from the Sun's harmful rays. People with blue eyes are more sensitive to sunlight and bright lighting because their eyes have less protection.

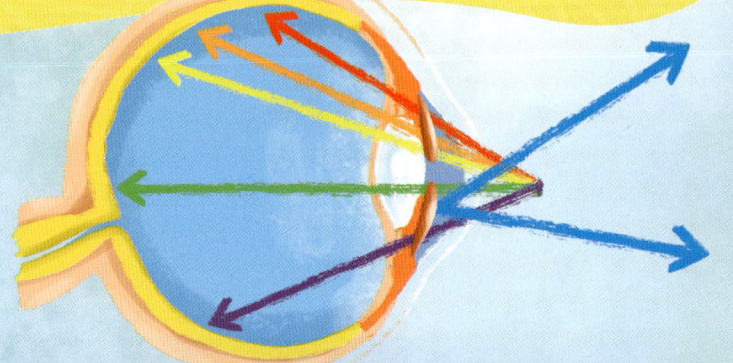

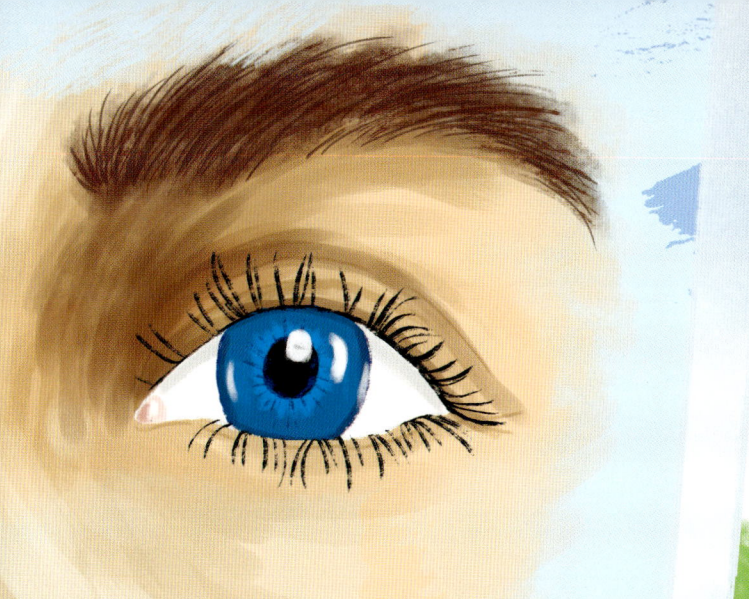

Genes are chains of instructions that build your body and make you unique. They determine things like your height, body shape, and hair and eye colour. You inherit genes from your biological parents.

MUTINY IN THE BODY

Sometimes a gene can be broken or missing. This is called a mutant (or mutated) gene, and can be passed down from a parent to their children.

NEW TO BLUE

Around 10,000 years ago, someone was born with a genetic mutation that meant they had blue eyes. This mutant gene was passed down through generations. Everybody on Earth today who has blue eyes (about 560 million people) is related to that first blue-eyed ancestor!

HOW RARE ARE YOU?

• Green is the rarest eye colour in the world.

• Less than 2 per cent of people on the planet have red hair.

• The rarest hair and eye colour combination is red hair and blue eyes.

DID YOU KNOW?

Some people have a rare condition called heterochromia, where their eyes are two different colours. Animals can have it, too.

COLOURS AND FEELINGS

Colours can have a powerful effect on the body and brain. Different shades change how we feel, think and behave. Discover how colour can make you feel lazy or lively, cool or creative!

GREEN
Slows our breathing and our heartbeat.

Makes us feel safe and secure.

Relaxes the body.

BLUE
Helps us think clearly.

Slows our breathing and heartbeat.

Soothes and cools the body.

RED
Speeds up our reactions and makes us more alert.

Increases our heart rate.

Speeds up our breathing.

Makes us feel warmer.

YELLOW
Gives our memory a boost and helps us think clearly.

Makes us feel positive and cheerful.

Makes us feel energetic.

Makes us feel warm.

PURPLE

Sparks the imagination and makes us feel creative.

Calms the body.

ORANGE

Makes us feel sociable.

Increases our heart rate.

Gives us an appetite.

A COLOURFUL LIFE

Red is often used inside cafés and bars because it is said to make us feel thirsty so we buy more drinks.

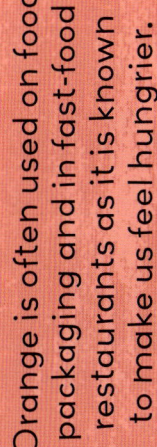

Orange is often used on food packaging and in fast-food restaurants as it is known to make us feel hungrier.

Surgeons often dress in green or blue in order to help calm and relax patients.

Studies show that installing blue lights in train stations or streets can reduce crime.

Cool colours like blue and lilac are often used to decorate offices because they're said to make people more productive and creative.

Green is the easiest colour for the brain to process. We can see more shades of green than any other colour.

Yellow helps us make decisions and act quickly. It is often used in fast-food chains so that we choose swiftly and move on.

DID YOU KNOW?

In one study, students who had to wait in a brightly coloured, red room before taking an exam got much lower test results than students who waited in a blue room, as it was more distracting.

BIG QUESTIONS ABOUT THE BODY

Find the icky answers to lots of colourful questions about the insides and outsides of your body.

WHY IS BLOOD RED?

Blood contains red blood cells that carry oxygen around the body in blood vessels (veins and arteries). Red blood cells have a protein called haemoglobin that is full of iron. When iron mixes with oxygen it turns red.

Blood that has been to the lungs for a fresh supply of oxygen (oxygenated blood) is bright red.

Blood that has already delivered its oxygen (deoxygenated blood) is still red, but it is much darker.

WHY ARE VEINS BLUE?

If you shine white light (a mixture of all light's different coloured wavelengths) onto your arm, your veins appear blue.

This is because red light travels in long waves, and passes through the skin, muscle and fat easily before being absorbed by the haemoglobin in the blood. Blue light travels in shorter waves and is mostly reflected by the skin. This means that the light returning to your eyes will contain more blue than red, making the veins look blue. It is a similar effect to how the Sun appears red at sunset. (Read about that on page 15.)

DID YOU KNOW?

Medics sometimes shine a red light on the skin to help them find a vein when taking blood or giving an injection.

WHY DO BRUISES CHANGE COLOUR?

When you bruise yourself, you break tiny blood vessels under the skin. The blood leaks out and pools around the area.

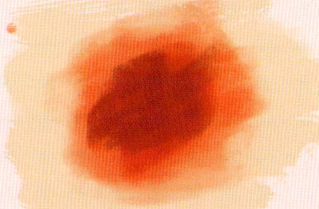

1 days
The bruise is red because the leaked blood contains oxygen.

2-5 days
The bruise darkens because the leaked blood is no longer being supplied with oxygen.

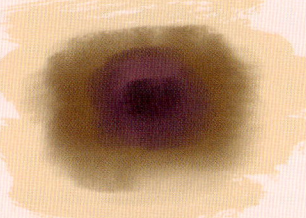

5-10 days
The body needs to clean up the spilt blood. It starts to break down the haemoglobin, and turns green then yellow.

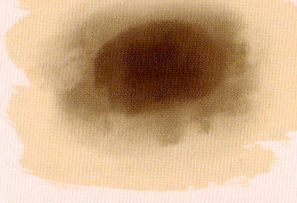

10-14 days
The healing has finished and the bruise fades.

WHY IS SNOT GREEN?

The slimy snot in your nose is called mucus. Together with the hairs in your nostrils, it traps any dust, dirt and germs that you breathe in and stops them travelling further into your body. We cough or sneeze out the unwanted invaders in our snot.

Gross green or yellow mucus shows that the body is fighting an infection. The colour comes from the white blood cells that rush to kill the offending germs. Once the cells have done their work, they die and are discarded in your snot, giving it its sickly tinge.

WHY DO WE TURN RED WHEN WE ARE EMBARRASSED?

Oops! Our face flushes with embarrassment and we can't control it! The redness is caused by veins in the face dilating (opening up), and more blood flowing into the cheeks. Scientists think this may be a kind of defence mechanism to avoid trouble, just like how dogs roll over to expose their bellies and cats flatten their ears to show that they are not looking for a fight. Us humans show that we are sorry and mean no offence with a bright red face!

DID YOU KNOW?

Some types of octopus, squid, spider and crab have blue blood. Instead of iron-filled haemoglobin, their blood contains copper. When copper mixes with oxygen, it turns blue.

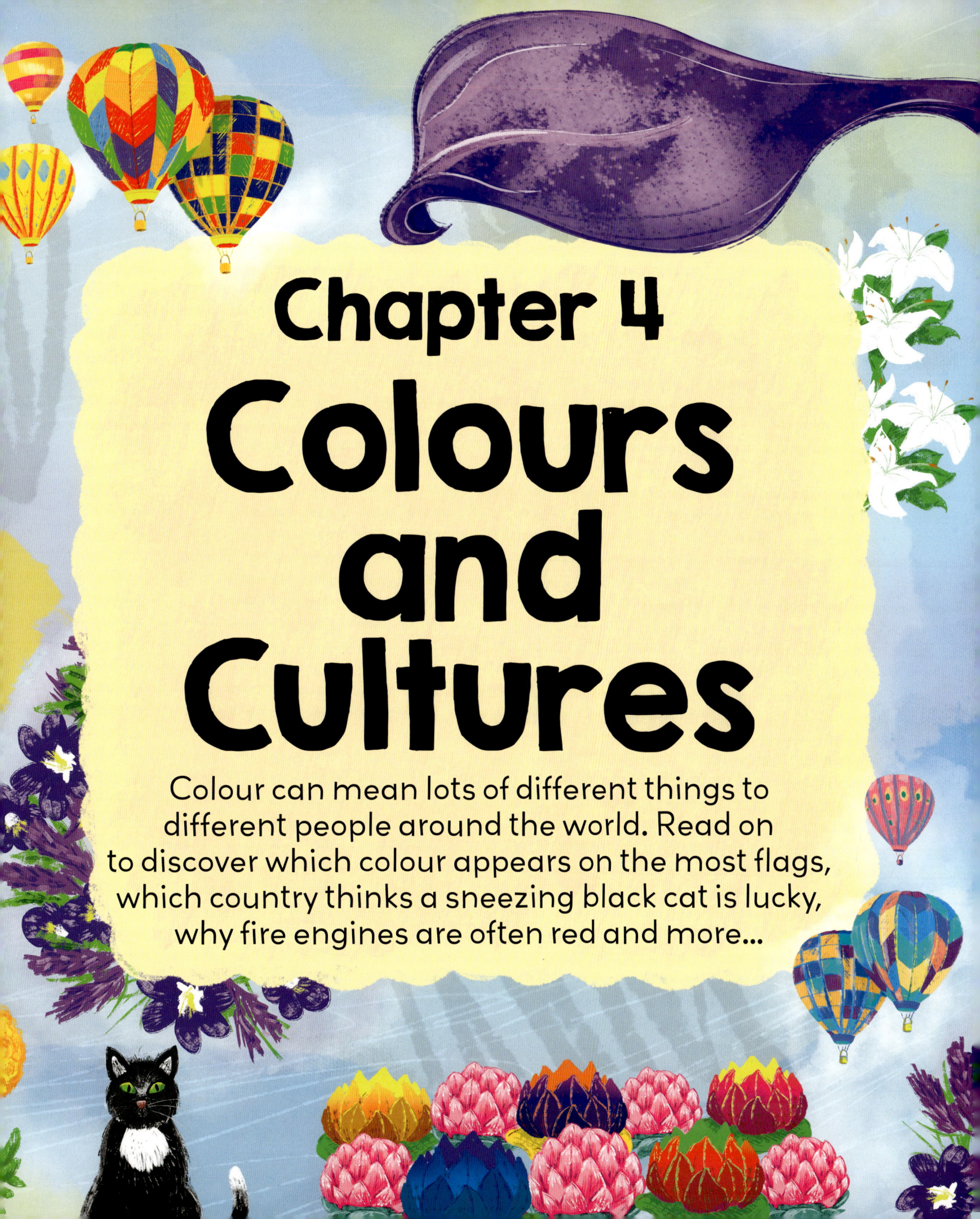

Chapter 4
Colours and Cultures

Colour can mean lots of different things to different people around the world. Read on to discover which colour appears on the most flags, which country thinks a sneezing black cat is lucky, why fire engines are often red and more...

EYE-POPPING COLOURS

Did you know that colours can transform right before your eyes? They can go from fading into the background to glowing brightly without actually changing at all!

THE COLOUR WHEEL

The primary colours are yellow, blue and red. They can be mixed to make all the other colours of the rainbow.

Colours that are close together on the colour wheel are similar, and colours that are far apart are different. Colours that are opposite each other on the colour wheel are as different as they can be. These are called **complementary colours.**

Blue is opposite orange:

Red is opposite green:

Yellow is opposite purple:

WHAT A CONTRAST!

Complementary colours have a strong contrast. For example, blue next to purple does not have a strong contrast as the colours are similar and the blue looks dark. On the other hand, blue next to orange has a strong contrast, since the blue looks brighter when it's next to its complementary colour. It is the same blue, but our eyes see it differently.

In 1861, a French chemist called Michel Eugène Chevreul made a discovery about how we see colour. Customers at the Paris tapestry company where he worked wanted more colourful artwork, so Chevreul was given the job of making brighter dyes.

During his work, Chevreul realised that the dyes didn't need to change; the colours just needed to be combined differently. Chevreul created a colour wheel to explain his ideas.

DRAMATIC WORK

Many artists were inspired by Chevreul's findings and began to create eye-catching art with complementary colours.

Impressionist painters used dots and splodges of complementary colours, not worrying that these were not the real colours of the subjects they were painting. From a distance, our eyes combine the different shades and make the colours look brighter.

Zoom in on this painting by Georges Seurat and you can see the blobs of contrasting colour.

Other painters used bold brush strokes and solid blocks of contrasting colours to make an impact:

Starry Night, Vincent Van Gogh, 1889

Woman with a Watch, Picasso, 1932

The Cat with Red Fish, Henri Matisse, 1938

BRIGHT BRANDS

Today, complementary colours are often used in well-known brands to make them stand out. Can you think of any?

CREATING COLOURS

Today, we can choose paint, pens and clothes in any colour we want. In the past, though, making things colourful was a lot more complicated.

Many modern colours are created from chemicals and made in big factories, but until a few hundred years ago, people had to make colours by hand. It was a long, difficult and sometimes dangerous process.

EARTHY SHADES

Tens of thousands of years ago, artists painted pictures on cave walls. They made their paint using whatever they had to hand.

RECIPE FOR PREHISTORIC PAINT

1. Grind up different coloured dirt to make red, orange, yellow and brown.
2. Use burnt wood to make black, and chalk to make white.
3. Mix the powders with a liquid to make it stick to the walls – popular choices include vegetable juice, wee, tree sap and animal fat or blood! Don't do this in your living room...

RECIPE FOR EGYPTIAN BLUE PAINT

1. Take desert sand, crushed limestone, and salt from a dry riverbed.
2. Add in shavings of copper or bronze and mix together.
3. Roll the mixture into small balls and bake at 1,000 °C.
4. Grind the baked balls into blue powder paint ready to decorate the royal tomb.

TRUE BLUE

The colour blue had special meaning to the ancient Egyptians. They invented a blue pigment that was used to paint royal tombs and coffins, and to decorate temples to their gods.

RECIPE FOR TYRIAN PURPLE DYE

1. Collect thousands of snails from the Mediterranean Sea.
2. Crack the shells and remove the snails.
3. Soak in salty water, then boil for 10 days. WARNING – this is smelly!
4. Dunk your chosen clothing and show off your fancy outfit to the townspeople.

RICH PURPLE

Purple was first made over 2,000 years ago. The dye came from sea-snail slime! It takes around 10,000 snails to make 1 g of dye, so it was an expensive process. Only the richest people could afford to use it, which is why purple was the colour worn by emperors, priests and kings.

DEATHLY WHITE

For over 2,000 years, lead was used to create white paint and make-up. In the 16th century, posh people in England used the poisonous metal as make-up to whiten their skin, as a pale complexion was a sign of youth and nobility – it meant you hadn't had to work in the fields all day. Over time, though, the lead caused hair loss and rotted teeth, so don't try this at home!

RECIPE FOR ELIZABETHAN MAKE-UP

1. Grind up a lump of lead, releasing its deadly poison.
2. Mix the powder with vinegar to form a paste.
3. Smear it over your face to create perfectly smooth, pale skin.

RECIPE FOR RED DYE

1. Collect thousands of female cochineal beetles from prickly pear cacti.
2. Boil then dry the bugs.
3. Crush the bugs to make a red powder (70,000 beetles make 500 g of dye).
4. Use your dye to make your new red clothing!

JUICY RED

Tiny cochineal beetles from South America make a bright red acid to deter hungry ants and birds. The beetles were used to dye cloth for thousands of years.

DID YOU KNOW?

Cochineal powder is still used to colour make-up and food. If the labels says "carmine", "cochineal extract", "E 120" or "natural red 4", it's actually beetle juice!

COLOUR CODES

Sometimes colours can communicate a message better than words or sounds. Colour codes give an instant warning or a signal that everybody understands immediately.

RED

Red signs, lights and buttons are often associated with danger.

They tell us to beware of potential hazards to prevent accidents.

WHAT DOES RED MEAN?

Danger	Warning	Wrong	Stop

GREEN

Green signs, lights and buttons are often associated with safety.

They lead us to safety, give safety information or tell us that we are safe.

WHAT DOES GREEN MEAN?

Safety	Working	Correct	Go

TRAFFIC LIGHTS

In every country where lights are used to control traffic, the colour code is the same.

RED LIGHT

As well as sending a warning, red light travels the greatest distance. Of all the colours, red has the longest wavelength, which means it can pass through particles in the atmosphere without scattering. Whether there is fog, snow, dust or smoke in the air, red light shines through to our eyes.

Brake and taillights on cars are red for the same reason. They are visible from a distance, warning other drivers that there is a car in front.

WHY ARE FIRE ENGINES RED?

All over the world, when there's a fire or an emergency, it's usually a red truck that you see racing to the rescue - but why?

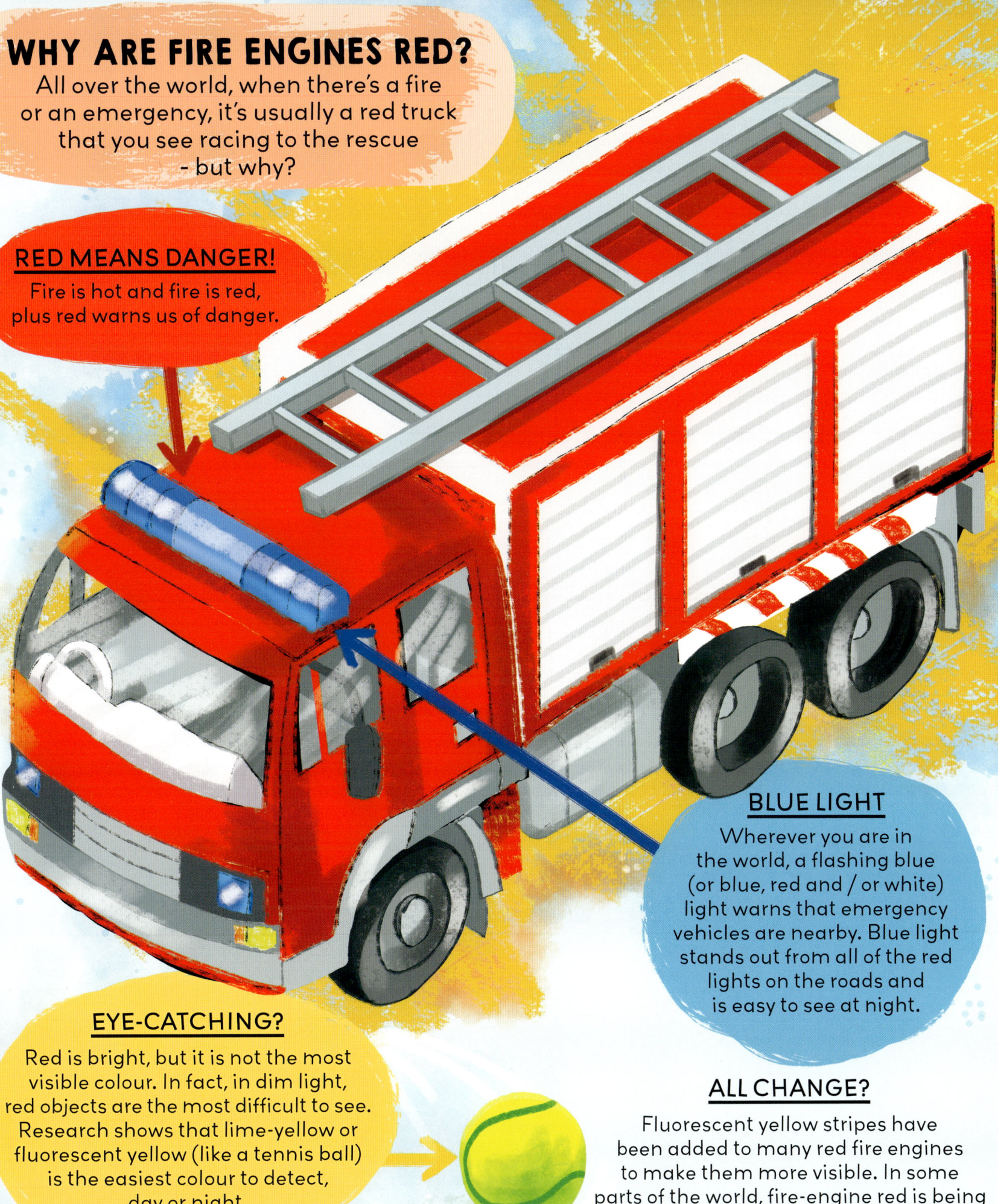

RED MEANS DANGER!
Fire is hot and fire is red, plus red warns us of danger.

BLUE LIGHT
Wherever you are in the world, a flashing blue (or blue, red and / or white) light warns that emergency vehicles are nearby. Blue light stands out from all of the red lights on the roads and is easy to see at night.

EYE-CATCHING?
Red is bright, but it is not the most visible colour. In fact, in dim light, red objects are the most difficult to see. Research shows that lime-yellow or fluorescent yellow (like a tennis ball) is the easiest colour to detect, day or night.

ALL CHANGE?
Fluorescent yellow stripes have been added to many red fire engines to make them more visible. In some parts of the world, fire-engine red is being swapped for day-glow yellow.

FLAGS OF THE WORLD

A nation's flag is much more than just a piece of colourful cloth. Its patterns, symbols and colours can sum up the dreams and identity of a whole country!

FRIEND OR FOE

It first became important for national flags to be bright and bold during times of war. It was hard to see much amongst charging horses and smoking guns on the battlefield, or thick fog and spray at sea. An eye-catching flag let everyone know who was who.

PICK YOUR COLOURS

• Red is the most common colour, appearing on 74 per cent of national flags.

• Red, white and blue is the most common colour combination, used by around 15 per cent of the countries in the world.

• Purple is the rarest flag colour. It appears on just three national flags – Dominica, El Salvador and Nicaragua – and only in tiny amounts!

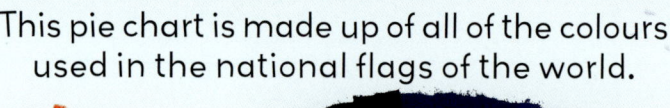

This pie chart is made up of all of the colours used in the national flags of the world.

WHAT IS THE LEAST COLOURFUL FLAG?

The Japanese flag is unique and instantly recognisable. Its red circle represents the sun, as the Japanese name for the country (*Nihon* or *Nippon*) means "where the sun rises".

TRICOLOUR

Bands of three colours, known as "tricolour", are found in more than 100 national flags. These stripes are simple and stand out at a distance.

Italy   France   Netherlands

DID YOU KNOW?

The Netherlands flag is the oldest tricolour. Its colours and design have inspired the flags of many other countries.

Red represents: strength, power, bravery, blood, passion and revolution.

Blue represents: water and sky, wisdom, truth, harmony, justice and prosperity.

Green represents: nature, rebirth, fertility, peace, growth, hope and the Islamic faith.

Yellow represents: wealth, courage, power, sunshine, independence and happiness.

WHAT DOES IT ALL MEAN?

Flag colours can reflect an old country's history or a new country's hopes for the future. Some colours have universal meaning, while others mean different things to different countries.

SAME BUT DIFFERENT

Green: the Catholic people

White: the hope of peace between people

Orange: the Protestant people

Ireland   Ivory Coast

Orange: the savanna and national growth

White: peace and unity

Green: the lush jungle and hope for the future

Some flags have very similar colours and patterns, but very different meanings.

COLOURFUL STORIES

Many flags tell a nation's story through their colours. The UK's Union Flag, for example, is a combination of the flags of three united countries.

Many African countries were once ruled by European ones. When they were able to rule for themselves, they designed new flags. Many were inspired by Ethiopia, an independent African nation, and copied its flag colours.

Others combined the Ethiopian flag with the flag of their old European rulers.

The Ethiopian flag + The French flag = The Central African Republic flag

TRY IT!

How quickly can you name each of these national flags?

ANSWERS: 1. Switzerland 2. Brazil 3. Canada 4. Germany

SPECIAL COLOURS

RED

In China, red is believed to bring good fortune, so brides get married in red.

At Chinese New Year, people dress in red, hang red lanterns and children receive money in red envelopes.

YELLOW

In Canada and the United States, families tie yellow ribbons around trees to remember loved ones who are missing or have gone to war.

In China, yellow is the colour of wisdom, happiness and glory.
In the past, the emperor was the only person allowed to wear bright yellow.

BLUE

In many cultures, blue is a masculine colour and represents the birth of a boy. In others, blue is considered more feminine and represents the birth of a baby girl.

In Greece, blue and white are the colours of the country's flag – and its houses. Doors and shutters were first painted blue by fishermen, as it was the cheapest paint available. Now it has become a colourful tradition.

ORANGE

In the Netherlands, people celebrate sporting events and the king's birthday with *Oranjegekte* ("orange madness"). Streets are decorated with orange flags, and everyone wears orange clothes and hats.

In Mexico, orange marigolds are used to decorate altars to the dead. The strong fragrance and colour of the flowers are thought to guide spirits back home.

Colours are connected with customs, celebrations and traditions around the world. They can be lucky or unlucky, or hold a special meaning. Joy, sadness, birth and death are all communicated with colour.

PURPLE

Purple is the colour of death and mourning in Brazil. Many of the country's Catholics believe that it is unlucky to wear purple at any other time. In Thailand, widows wear purple.

BLACK

In Europe and North America, black is the colour of death and mourning.

In Asia, black is a symbol of experience and skill. The black belt is the highest level that can be achieved in some martial arts.

In the UK, you're in luck if a black cat crosses your path. You'll find love if you spot one in Japan, whereas a sneezing black cat will bring good fortune in Italy.

GREEN

Green is the national colour of Ireland, which is known as the Emerald Isle because of its lush, green countryside.

DID YOU KNOW?

According to Irish legend, a leprechaun is a tiny, mischievous fairy. It's said they pinch anyone not wearing green on St. Patrick's Day (a celebration of the country's saint).

WHITE

In Western culture, white symbolises innocence. Brides traditionally wear white.

In India and parts of Asia, white is a symbol of death. White is worn to Hindu funerals to show grief for a loved one.

COLOURFUL CELEBRATIONS

Celebrations and festivals take place every day of the year across the globe. These are some of the most beautiful, colourful (and sometimes messiest) sights on Earth!

INTERNATIONAL BALLOON FIESTA, ALBUQUERQUE

Albuquerque, in New Mexico, USA, holds the world's biggest balloon festival. For nine days each October, the dusty desert becomes a sea of colour as hundreds of hot-air balloons inflate, then drift silently into the blue sky.

LA TOMATINA, SPAIN

This huge, squelchy festival takes place in the town of Buñol in Spain. Every August, thousands of people take part in the world's biggest food fight. Around 150,000 kg of over-ripe tomatoes are dumped in the streets for people to toss at each other. For a short time, the town and the people turn a juicy shade of red.

HOLI, INDIA

Every spring, people across India and around the world celebrate Holi, the Hindu Festival of Colours. This vibrant festival involves throwing brightly coloured powders at one another to welcome springtime, and celebrate colour and victory over evil.

THE LOTUS LANTERN FESTIVAL, SOUTH KOREA

Known as *Yeon Deung Hoe* ("The Feast of Lanterns") in Korean, the Lotus Lantern Festival celebrates the Buddha's birthday. (Buddha began the Buddhist religion over 2,000 years ago.) Hundreds of thousands of colourful, candlelit lanterns are raised throughout the streets and temples. The paper lanterns are shaped like lotus flowers, a symbol of purity and wisdom.

L'INFIORATA, GENZANO, ITALY

In May and June, some Italian towns hold *L'Infiorata*, which means "decorated with flowers". The streets are turned into giant artworks using flowers, seeds and petals. In Genzano, it takes a team of artists and half a million blooms to create the colourful flower carpet. After two days, children are allowed to run down the street, undoing all of the hard work.

RIO CARNIVAL, BRAZIL

The Rio Carnival takes place in the Brazilian city of Rio de Janeiro, and is known as the world's biggest party. Two million people crowd the streets each day to watch colourful parades of floats and thousands of dancers in incredible costumes.

COLOURFUL LANGUAGE

Many languages have colourful expressions to describe everyday life. The feelings and situations are the same all over the world, but the colours can often be very different!

Mmm! This looks delicious!

LIES

Dishonesty comes in a range of colours depending on which language you're speaking.

In English, **a white lie** is one that is told to be polite or avoid hurting someone's feelings.

In Turkish, **a pink lie** is a small or harmless lie.

In Korean, **a red lie** is one which everyone knows is a lie.

In German, if you **lie the blue out of the sky**, you lie constantly!

ANGER

It turns out rage can be a rainbow of colours.

In English, **seeing red** is when someone becomes suddenly angry.

In Romanian, you **turn pink from rage.**

In Polish, you **turn purple in the face.**

In Thai, Greek and Italian, you **turn green with anger.**

In German, you can be **annoyed green and blue.**

In French, **being white-hot** is to be incredibly angry.

In Finnish, if you are said to **turn black,** you are really raging!

JEALOUSY

This unpleasant feeling has some enviable shades.

In English, you might **be green with envy** or even a **green-eyed monster**.

In French, Hungarian, German and Italian, you can **be yellow with jealousy**.

In Swedish, a jealous person is said to **have a black disease**.

HAVING NO MONEY

Languages are rich with ways to describe being poor.

In Italian, when you are broke, you are **at the green**.

In Spanish, if you run out of cash you are **without white**.

In English, when your bank account is empty you are **in the red** (or **in the black** if you have lots of money!).

In Dutch, if you have little or no money, **you sit on a black seed**. This expression might come from birds leaving the less-tasty dark seeds until last.

LOVE LANGUAGE

Love isn't just about red hearts and roses – it can be other lovely colours too!

In Swedish, Danish and Dutch, loved-up people are said to **float on pink clouds**.

In Belgian French, being in love is **being blue** for somebody.

In Brazilian Portuguese, **to have seen the green bird** is when a person is smiling because they're in love.

TRY IT!

CAN YOU GUESS WHAT THESE COLOURFUL EXPRESSIONS MEAN?

1. POLISH:
"Think about blue almonds!"

2. SPANISH:
"You are giving me green hairs!"

3. GREEK:
"Green horses!"

4. ENGLISH
"Paint the town red!"

ANSWERS: 1. To daydream. 2. You are annoying me. 3. Expression of disbelief. 4. To go out and enjoy yourself.

INDEX

A
Anger 62
Animals 22–31
　displays 22–23, 26
Ants 24
Art 51–53

B
Baby animals 27
Bees 31, 34
Beetles 25, 26, 53
Berries 34
Birds 23, 26–27, 34–35
Black 24, 38, 59
Blood 46–47, 52, 57
Blue 44, 52, 58
　Earth 8, 14
　feelings 11, 55
　water 17
Brown 9, 52
Bruises 47
Bulls 30
Butterflies 22, 25, 31, 34

C
Camouflage 27, 28, 35
Caterpillars 25
Celebrations 58–59, 60–61
Chevreul, Michel Eugène 51
Chlorophyll 32–33
Clouds 9, 10, 16
Colour blindness 39
Colour wheel 50–51
Customs 58–59, 62–63

D
Danger 24–25, 54–55
Deer 30

E
Earth 8–9, 14, 15, 18–19
Egypt, ancient 52
Emotions 26, 44–45, 62–63
Eye colour 42–43
Eyesight 30–31, 38–39

F
Feelings 26, 44–45, 62–63
Festivals 58–59, 60–61
Fire engines 55
Flags 56–57
Flowers 31, 34–35, 58, 61
Frogs 25
Fur 28–29

G
Gecko 31
Genes 43
Grass 32–33
Green 8, 44, 59
　safety 54
Grey 9, 16

HIJ
Human bodies 38–47
Insects 24–26, 31, 34, 35
Jealousy 63

L
Lies 62
Lizards 23, 25
Love 63

M
Melanin 42
Money 58, 63
Monkeys 22
Moon 14

N
Newton, Isaac 12
Night vision 31, 39
Northern Lights 19

O
Octopuses 24, 26
Orange 45, 58

P
Paint 52–53, 58
Plants 31–35
Polar bears 28–29
Prehistoric art 52
Prisms 12
Purple 45, 53, 59

R
Rainbows 12–13
Red 11, 44, 53, 58
　danger 54–55
　sunsets 15

S
Sea 8, 17
Senses 40–41
　see also vision
Skinks 23, 25
Skunks 24
Slugs 25
Snakes 24, 25
Snot 47
Southern Lights 19
Space 10–11, 14–15
Spiders 22, 23, 27
Stars 10–11
Sun 14–15

T
Tigers 30
Toads 24
Traditions 58–59, 62–63
Traffic lights 54

UV
Ultraviolet light 14, 31
Veins 46–47
Vision 30–31, 38–39

W
Warnings 24–25, 41, 54–55
Water 8, 13, 15, 16–17
White 38, 53, 59
　Earth 8, 16
　polar bears 28–29
　space 11, 14–15

Y
Yellow 14, 44, 58